START WITH A QUESTION

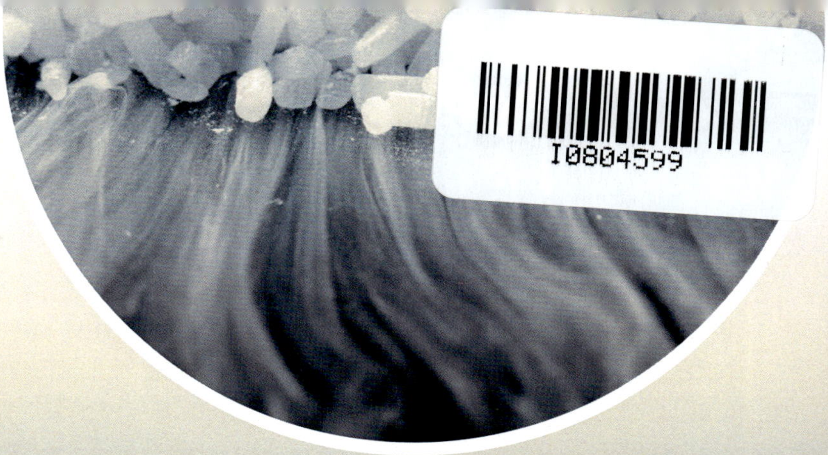

Is Peanut Butter a Liquid?

FAQs About Matter

Rachel Stuckey

A Crabtree Forest Book

Crabtree Publishing
crabtreebooks.com

Notes for Readers and Class Discussion

This book is designed to teach readers about core subject areas, build curiosity, and inspire further investigation. Readers are encouraged to build upon what they already know about the subject and engage in topics they want to learn more about. Here are a few guiding questions to prompt reflection and discussion. Possible answers appear in red.

Before Reading:

Read the title and look at the table of contents. What do you already know about matter?

- *I know that there are different forms, or states, of matter.*
- *I know that everything is made of matter.*

What would you like to learn about matter?

- *I would like to learn about unusual forms of matter.*
- *I would like to learn how matter changes state.*

During Reading:

Pause after reading each page or chapter. What questions do you have about what you read? What are you curious to know?

- *I wonder if there are even smaller pieces of matter than quarks.*
- *I am curious to know what caused the Big Bang.*

Make connections to what you are reading. How is this like something you already know?

- *I know that humans need to drink water to survive.*
- *I know that many products are molded from plastic.*

After Reading:

Recall key details about the book. What was the author trying to teach readers?

- *The author was trying to teach readers that matter is not created or destroyed—it only changes state.*
- *The author was trying to teach readers that everything in the universe is made of different forms of matter.*

How did the images and captions help you understand more?

- *The photographs helped me see different forms of matter in action.*
- *The captions helped me understand more details about matter.*

Table of Contents

THINK LIKE A SCIENTIST 4

CHAPTER 1
Questions About Matter 6

CHAPTER 2
Questions About Solids 14

CHAPTER 3
Questions About Liquids 22

CHAPTER 4
Questions About Gases 30

CHAPTER 5
Questions About Changing States of Matter 38

GLOSSARY 46
INDEX 47
COMPREHENSION QUESTIONS 47
ABOUT THE AUTHOR 48

Think Like a Scientist

Look around you. Do you wonder what makes up the ground beneath your feet, the air you breathe, or even the book you are reading? Scientists, too, asked questions about what made up the world around them. They carried out experiments and drew conclusions. They learned that everything in the universe is made of matter!

Matter is anything that has **mass** and takes up space. We measure matter in volume—how much space it takes up—and by density—how much matter is in that space. Over time, scientists discovered many facts about matter. It is made up of tiny, moving **particles** called atoms. It can exist in three main states: solid, liquid, and gas. A fourth state—plasma—is created when a gas is superheated. It is mostly found in stars, but also in lightning and neon signs.

Gas particles are far apart from each other and move very quickly. A gas will spread out to fill its container. It does not have volume.

Liquid particles are closer together and move fairly quickly. A liquid changes shape to fit its container. Unlike a gas, it has volume.

Solid particles are closely packed and do not move much. A solid has an unchanging shape and volume.

Matter can transform between states. Cold temperatures, for example, freeze water into ice.

Science is always changing as scientists ask more questions and discover new answers. Scientists learned that energy can change the state of matter. **Chemical reactions** can also change the composition, or makeup, of matter.

In this book, you can explore answers to questions you may have wondered about matter. The information you learn may inspire you to ask more questions—just as scientists throughout history have done.

Where does matter come from?

During the dark ages, there were no stars or other sources of light in the universe.

All of the matter in our universe was created from the Big Bang—a huge explosion that happened around 13.8 billion years ago. The cause of the Big Bang is unknown. Scientists do know that before the Big Bang, all of the matter and energy in the universe was in a tiny, extremely hot, **dense** point. When it suddenly exploded, it expanded rapidly to create the universe. Even today, the universe is still expanding outward from that point—but much more slowly than during the Big Bang.

Except for special conditions like star explosions, the matter in our universe is neither created nor destroyed—it just changes state. This is called the Law of Conservation of Matter. The most basic change is when we turn a solid into a liquid by melting it. We can also heat liquids to turn them into gases, such as when we boil water to create steam. Different types of matter can be combined to create new substances or materials. We also consume food to give us energy. But despite these changes, the number and type of atoms in our universe stay the same—and have existed since the Big Bang.

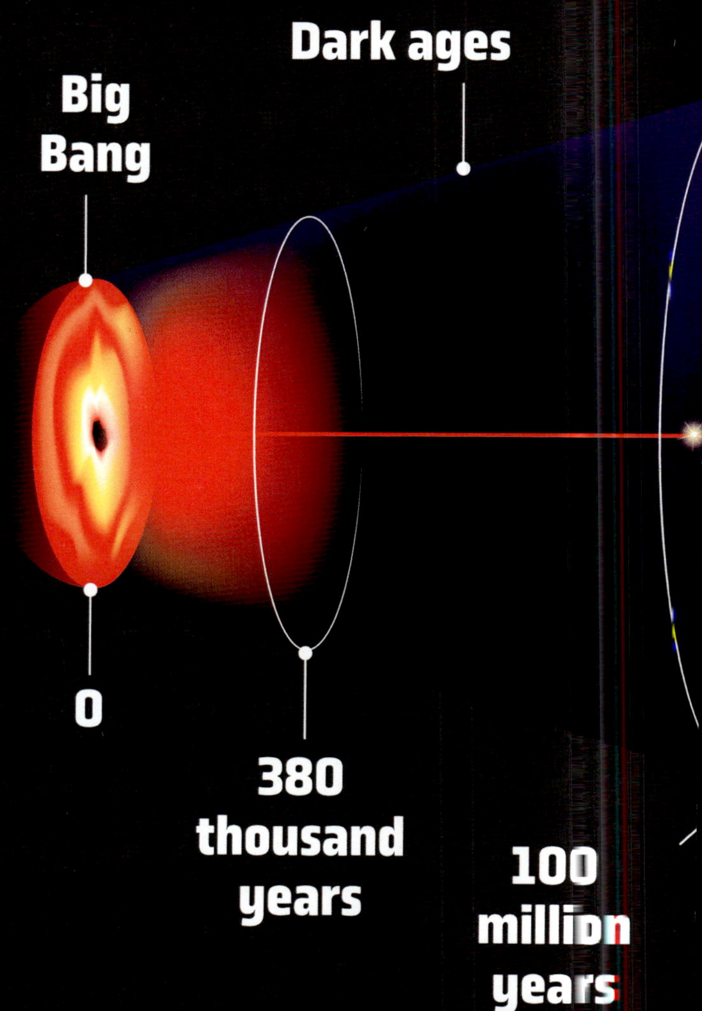

In the seconds and minutes after the Big Bang, the universe expanded and cooled down. This allowed energy to change into particles of matter.

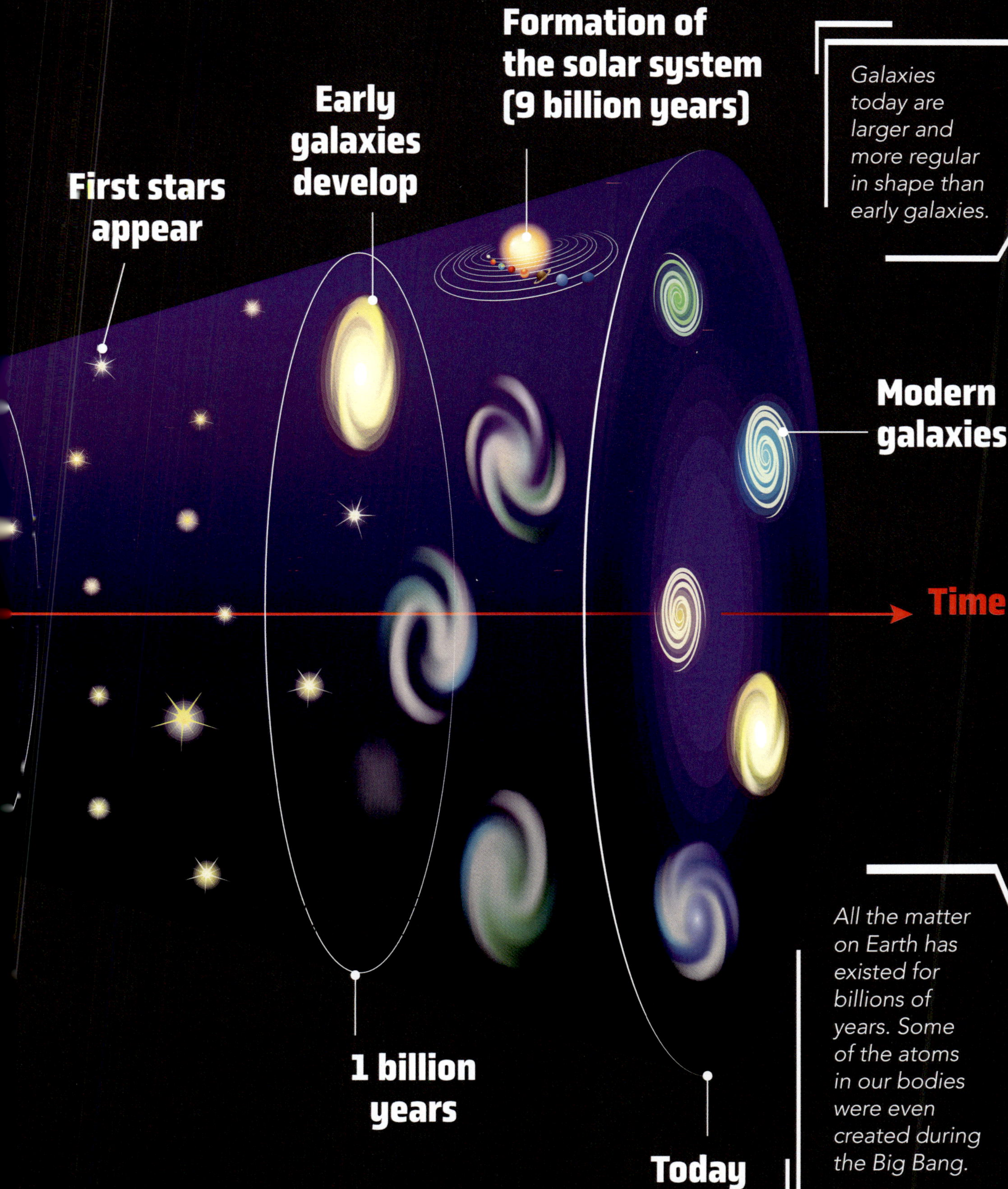

Galaxies today are larger and more regular in shape than early galaxies.

All the matter on Earth has existed for billions of years. Some of the atoms in our bodies were even created during the Big Bang.

Every type of matter is made of extremely tiny particles called atoms. Two or more atoms combine to create molecules. When matter is made up of one type of atom, it's called an element. An element cannot be broken down into other substances. Two or more different elements combine to create compounds. Water is a compound of two elements: hydrogen and oxygen. Each water molecule has two hydrogen atoms and one oxygen atom.

Water is one of the only substances on Earth that naturally exists as a solid, a water, and a gas. Most other substances can only be changed using human effort. For example, we can use heat to melt metals. Burning wood turns **carbon** solids into gas. But from ice in glaciers to water in lakes and water vapor in the air, water naturally exists in solid, liquid, and gas forms. It is constantly changing state as part of the **water cycle.**

Elements

Elements can be made of one atom, or of molecules with one type of atom.

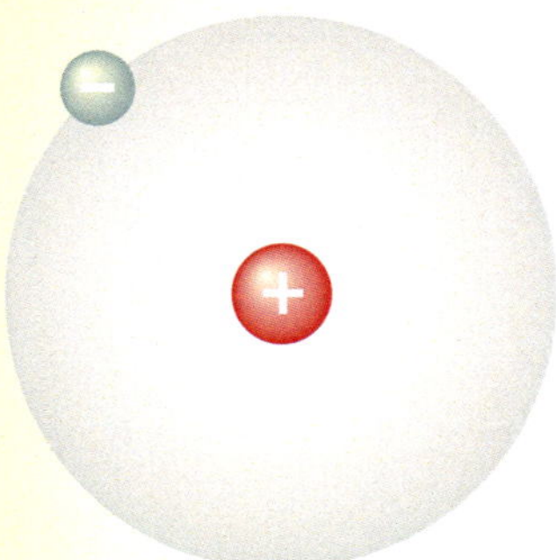

Hydrogen atom

Electron Proton

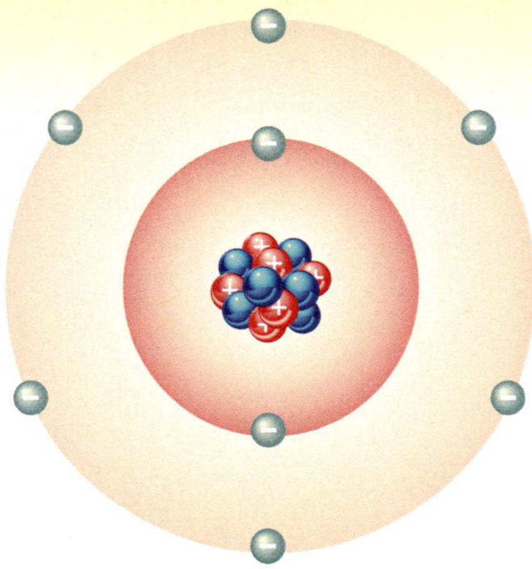

Oxygen atom

8 Electrons 8 Protons 8 Neutrons

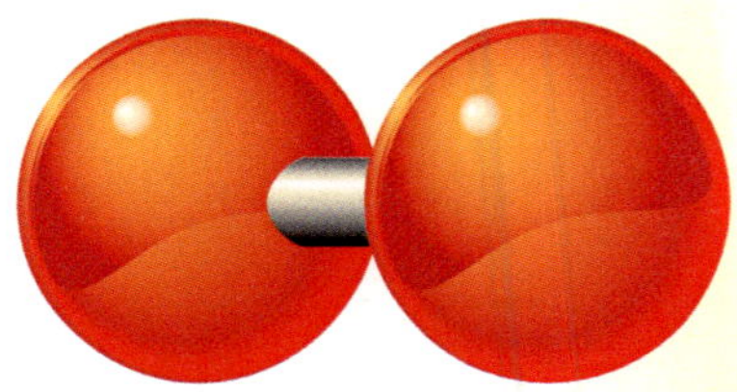

O_2

Oxygen molecule

2 oxygen atoms

Compounds

Compounds have molecules made of different types of atoms.

H_2O

Water molecule

2 hydrogen atoms + 1 oxygen atom

CO_2

Carbon dioxide molecule

1 carbon atom + 2 oxygen atoms

DID YOU KNOW?

Carbon is found in all living things. Like all matter, it is never created or destroyed. Instead, it moves through the carbon cycle. Plants use carbon from the air to make food through photosynthesis. Animals, including humans, eat the plants. Then they return carbon to the cycle through their waste. We breathe out carbon dioxide as a waste gas, for example. Carbon is also released into the air, soil, and water when living things die and decompose, or break down.

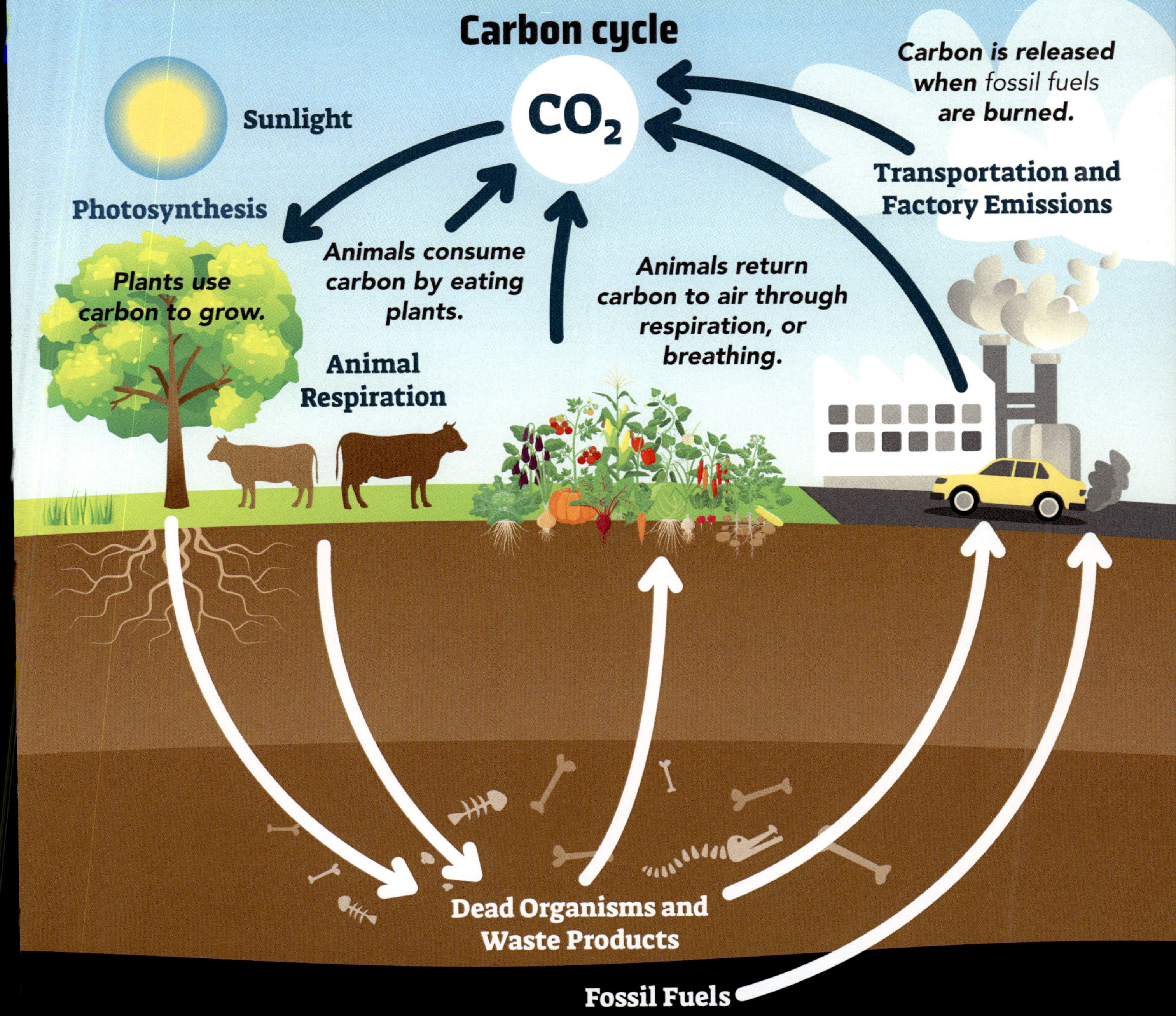

What is the smallest piece of matter?

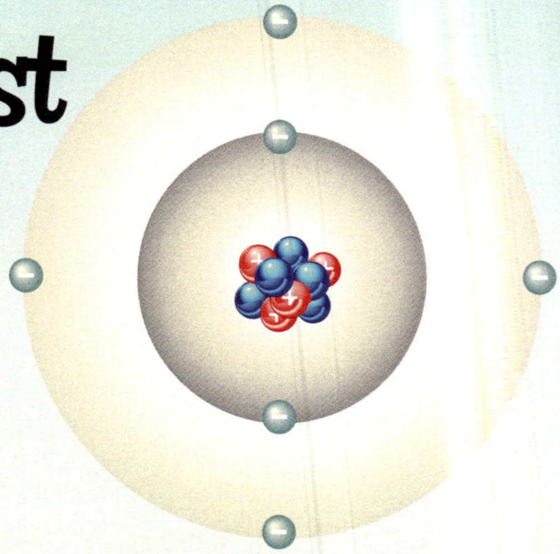

Carbon atom

6 Electrons

6 Protons

6 Neutrons

All matter is made of atoms. But not all atoms are the same size. Atoms are unique to each element. There are 118 different known elements, listed on the periodic table of elements. The elements—and therefore their atoms—increase in mass as you move down the table.

There are smaller pieces of matter than atoms, however. Every atom is made of three types of particles: neutrons, protons, and electrons. But still, there are even tinier particles. Scientists think the tiniest piece of matter is the quark. Protons and neutrons are made up of quarks.

The first element listed on the periodic table is hydrogen, which has an atomic number of 1. The atomic number is the number of protons in an atom's nucleus, or center.

Neutrons and protons clump together in the ***nucleus*** *of the atom. The much smaller electrons surround the nucleus.*

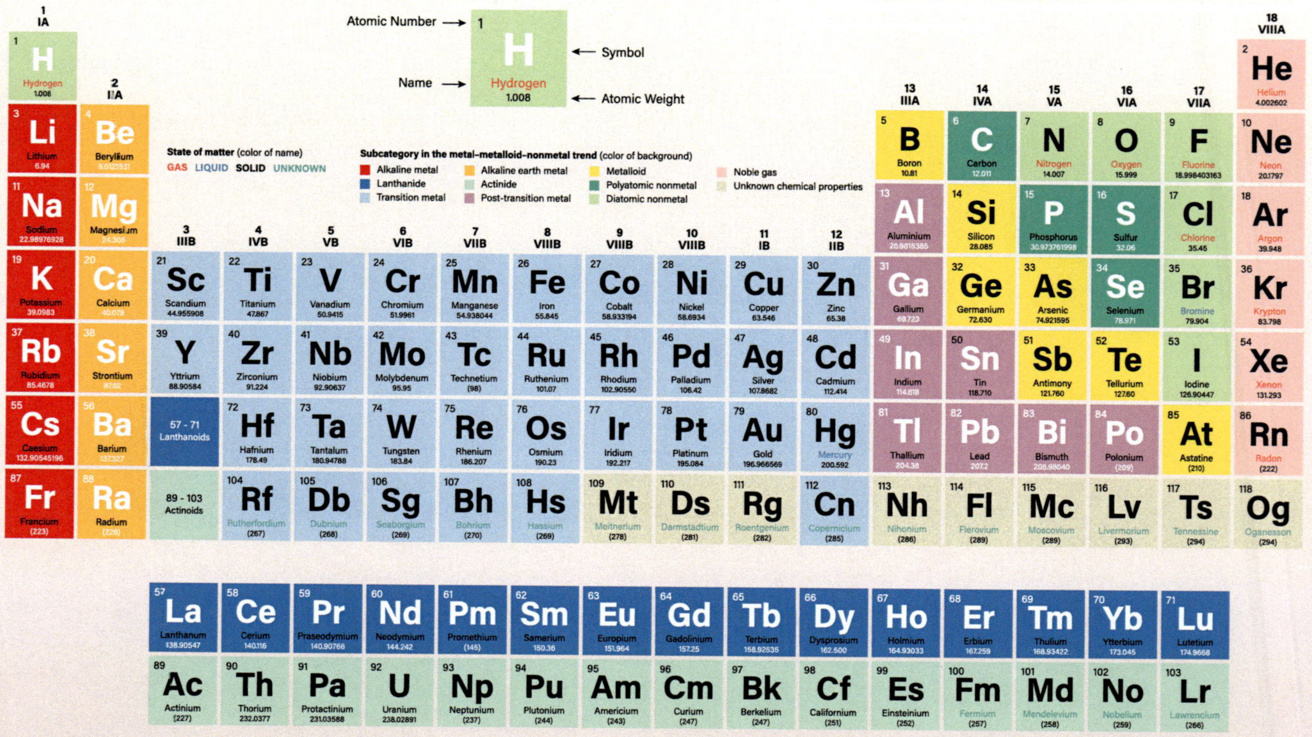

Can you touch a rainbow?

Rainbows are caused by light traveling through water droplets in the air. That means rainbows can only be seen, not touched. We can only see rainbows when the sunlight, conditions in the air, and our position on the ground are just right.

Light is made up of waves, which have different **wavelengths**. When the wavelengths are separated, each is a different color. Together, they look like white light. When light from the Sun hits water droplets in the air, the wavelengths separate into a circle of colored light.

We only see part of a rainbow's circle because the ground stops the rest. Sometimes, pilots can see full-circle rainbows from planes.

How much does matter weigh?

Imagine a loaf of bread and a cement block that are roughly the same size. The cement block has a higher density, or more particles of matter in the same amount of space. It therefore has a greater mass.

Mass is similar to weight, but it is not the same. Mass is a measure of the amount of matter in an object. Weight is a measure of the **force** of **gravity** on an object. While mass does not change anywhere in the universe, weight changes based on the strength of gravity. That's why a person weighs less on the Moon, with its weaker gravity, than they do on Earth. Scientists measure mass in pounds or kilograms, but measure weight in **newtons**.

Astronauts are weightless in space in part because the gravity there is very small. However, they still have the same mass as they do on Earth. That does not change.

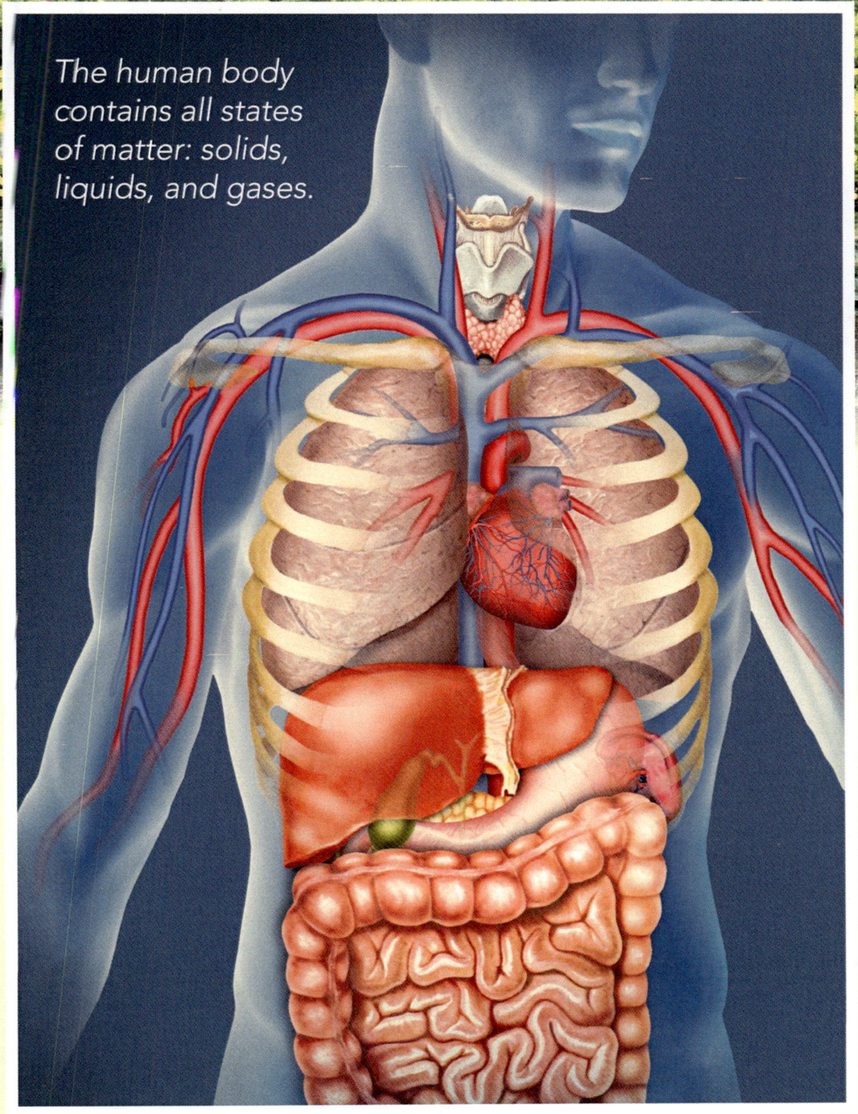

The human body contains all states of matter: solids, liquids, and gases.

1 IA	14 IVA
1 **H** Hydrogen 1.008	6 **C** Carbon 12.011
15 VA	**16 VIA**
7 **N** Nitrogen 14.007	8 **O** Oxygen 15.999

Hydrogen, carbon, nitrogen, and oxygen are found in all living things.

What types of matter are found in the human body?

The human body is made up of billions and billions of atoms. But just four elements make up around 96 percent of the body: oxygen, hydrogen, carbon, and nitrogen.

About 18 percent of the atoms in your body are carbon. Sixty-five percent are oxygen, and about 10 percent are hydrogen. Nitrogen is 3 percent. The rest is a mix of calcium, phosphorous, potassium, sulfur, chlorine, sodium, and magnesium. Tiny amounts of other elements, such as iron, are also found in the human body. These atoms combine into many types of molecules, such as **proteins**.

Why do rocks sink but wood floats?

Both rocks and wood are solids. Every rock you throw into the water—big or small—will sink. But a tree branch—big or small—will usually float. That's because of density. The reason rocks do not float is that they are denser than water. Wood, however, is less dense than water, so a tree branch floats on top.

The particles in all solids are tightly packed together. In hard solids, such as rocks, the force that holds the particles together is stronger than in soft solids, such as sponges. Some solids can be squeezed, bent, or even stretched. But all solids have a fixed shape that is difficult to change. We usually must use force to change their shape in some way.

*Eventually, wood will sink. That is because wood is a **porous** material. It will slowly take in water, increasing its density.*

DID YOU KNOW?

The solid form of matter is usually denser than its liquid form. But that isn't the case when it comes to water. Solid water—or ice—is less dense than liquid water. That is why ice floats—from ice cubes in your drink to icebergs in the ocean. This makes water a unique type of matter.

Ice is less dense than liquid water because of the way water molecules form a ***crystalline structure*** *when they freeze. The hydrogen atoms in this structure push the molecules farther apart, making the ice less dense.*

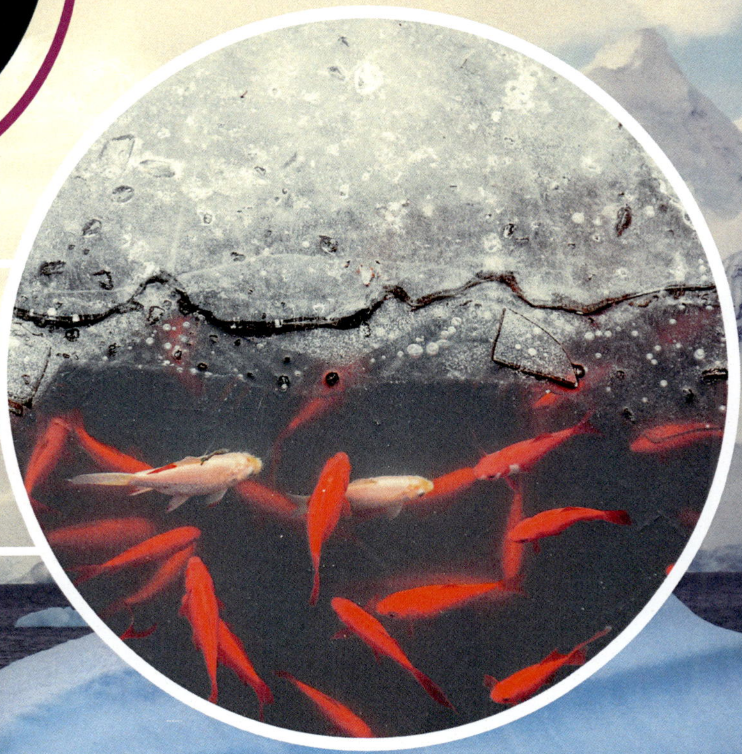

Plants and animals depend on floating ice to survive. Ice on the surface ***insulates*** *the water beneath, stopping it from freezing and protecting important water habitats.*

There are many different types of solid matter—not just rocks, wood, and ice. Each one is made of an element or a mix of elements. Material that is less dense than water will float. But sometimes the shape of the material makes a difference. A piece of metal will sink in water because metals are denser than water. But if you shape that metal into a boat, it floats because much of its shape now contains air—and air is less dense than water.

Solid matter can also range in hardness, which is different than density. The harder a substance, the more tightly packed its molecules. Some solids are very soft. Imagine a cotton ball. You can increase the density of a cotton ball when you squeeze it together. It may have more particles in less space, but the material is just as soft.

The hardest naturally occurring substance in the world is a form of carbon: the diamond. Its atoms are tightly bonded together. Diamonds are light enough to be worn as jewelry. But they are very difficult to break. In fact, diamonds are often used to cut and shape other diamonds.

A mineral's hardness is measured on the Mohs scale. Diamonds rank 10 out of 10.

Huge cruise ships carry a ton of weight. Their wide, ***U-shaped bodies,*** *however, make sure that they do not sink in water.*

What makes some solids bend and stretch, and others break?

Solids can be deformed in different ways. Some solids are brittle and will break when they are bent or pulled. Other solids are ductile, which means they will stay in one piece when pulled or bent. Materials such as glass are brittle, while many types of metals are ductile.

Some solids are elastic. Think of hair elastics or rubber gloves. These materials stretch when a force is applied, and then return to their original shape when the force is removed. But if too much force is applied, the connections between the molecules can be damaged.

The way that the molecules of solid matter attach to each other affects whether they are ductile, brittle, or elastic. These characteristics all exist in different degrees They can also change as matter starts to change its state. If you put a rubber band in the freezer, it will change from ductile to brittle and snap in two.

DID YOU KNOW?

Natural rubber is made from the sap of rubber trees. Indigenous **peoples in South America learned to scrape lines into the trees and collected the liquid that seeped out. This liquid, known as latex, is made up of small solid particles** suspended **in water. Acid is then used to pull out the solid rubber. Most natural rubber now comes from rubber trees in Southeast Asia. Artificial rubber, on the other hand, is made from oil products.**

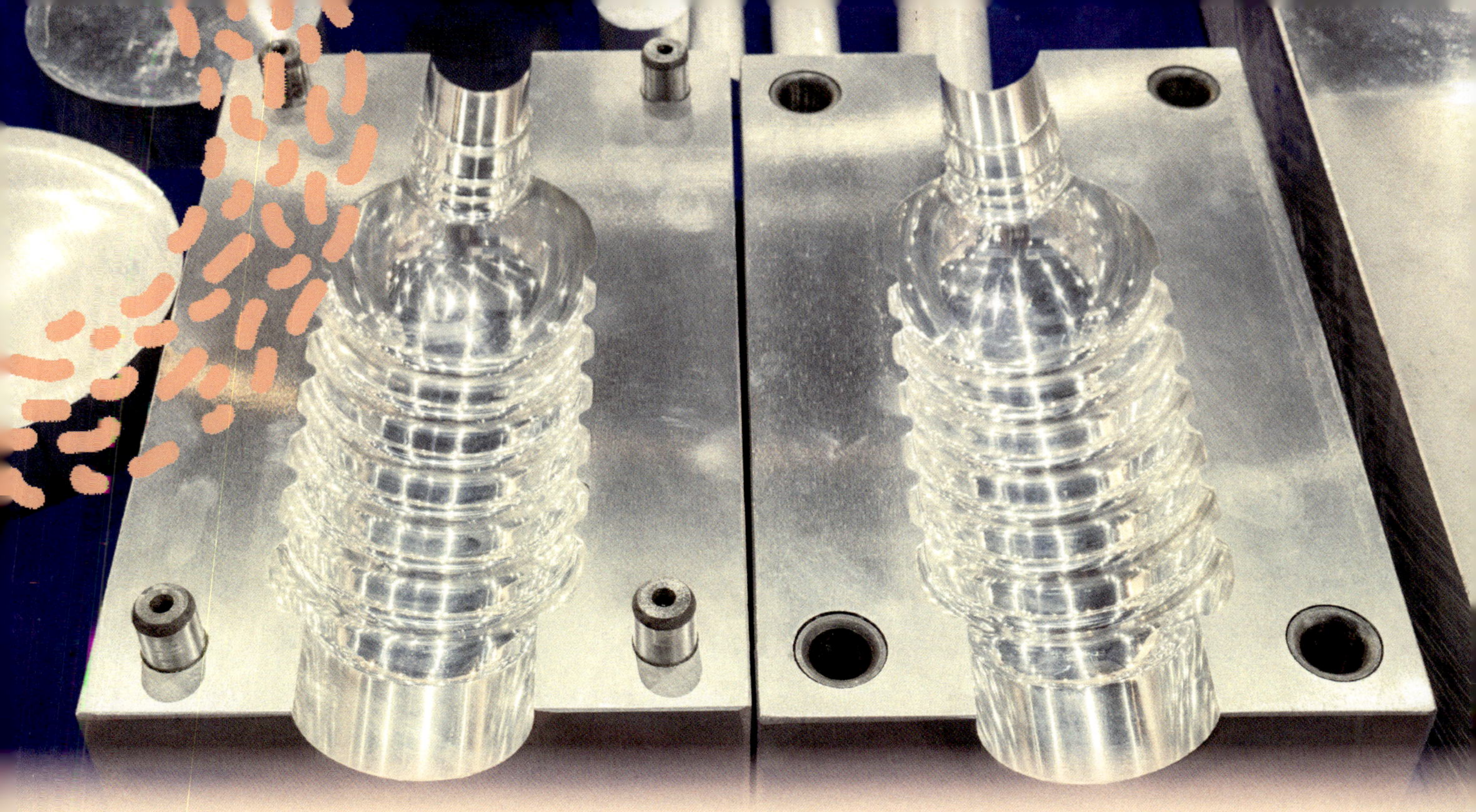

Plastic, a human-made solid, comes from a Greek word meaning "to mold or shape." In its liquid form, plastic is poured into molds, where it hardens into a solid.

At warmer temperatures, plastics can be ductile.

At lower temperatures, plastics are brittle.

Is snow a solid?

Yes, snow is a solid. It consists of a special kind of frozen water. When the temperature in the clouds is very low, water vapor freezes and becomes solid very quickly, skipping over the liquid stage. The water molecules form tiny ice crystals around tiny molecules of dust, and eventually fall to the ground as snow.

Crystals are a special type of solid with a crystalline structure. The molecules in this structure stick together very tightly in a regular geometric pattern. Some crystals can only be seen through a microscope, but others, such as snowflake crystals, can be seen in nature.

Sugar and salt are crystals in their solid form.

Though snow falls to the ground as a solid, if heated up it quickly turns to liquid form.

How is glass made?

Most glass is made by melting sand, limestone, and **soda ash**, then cooling the liquid quickly. If it cooled naturally, it would form a solid crystal. Cooling it quickly, however, creates a substance that we can see through. But the truth is, glass isn't a true solid!

Glass is an amorphous solid. It has some properties of a solid and some properties of a liquid. The molecules in glass are not tightly packed into a regular pattern, like a solid. However, they are also not moving as freely as they would in liquid form.

Amorphous solids are formed when a solid is melted and cooled very quickly.

What happens to the water when wet clothing dries?

When we hang laundry on a rack or clothesline, we know that eventually the wet clothes will dry. This happens through the process of **evaporation**. When liquid water heats up, the molecules start to separate and rise into the air as vapor. When water is boiled, evaporation happens quickly. At lower temperatures, such as the temperature at which you hang your wet clothes, the process happens much slower. The warmer the air, the faster your clothes will dry.

The evaporation of water is part of Earth's water cycle. Our **atmosphere** is full of water vapor. Most of it comes from evaporation from large bodies of water such as oceans, lakes, and rivers, but some of it comes from your wet clothes. As the temperature cools, **condensation** occurs. The water vapor becomes liquid again and returns to Earth as precipitation. The rain and snow return to bodies of water and to **groundwater**. And then, we can use this water to wash our clothes.

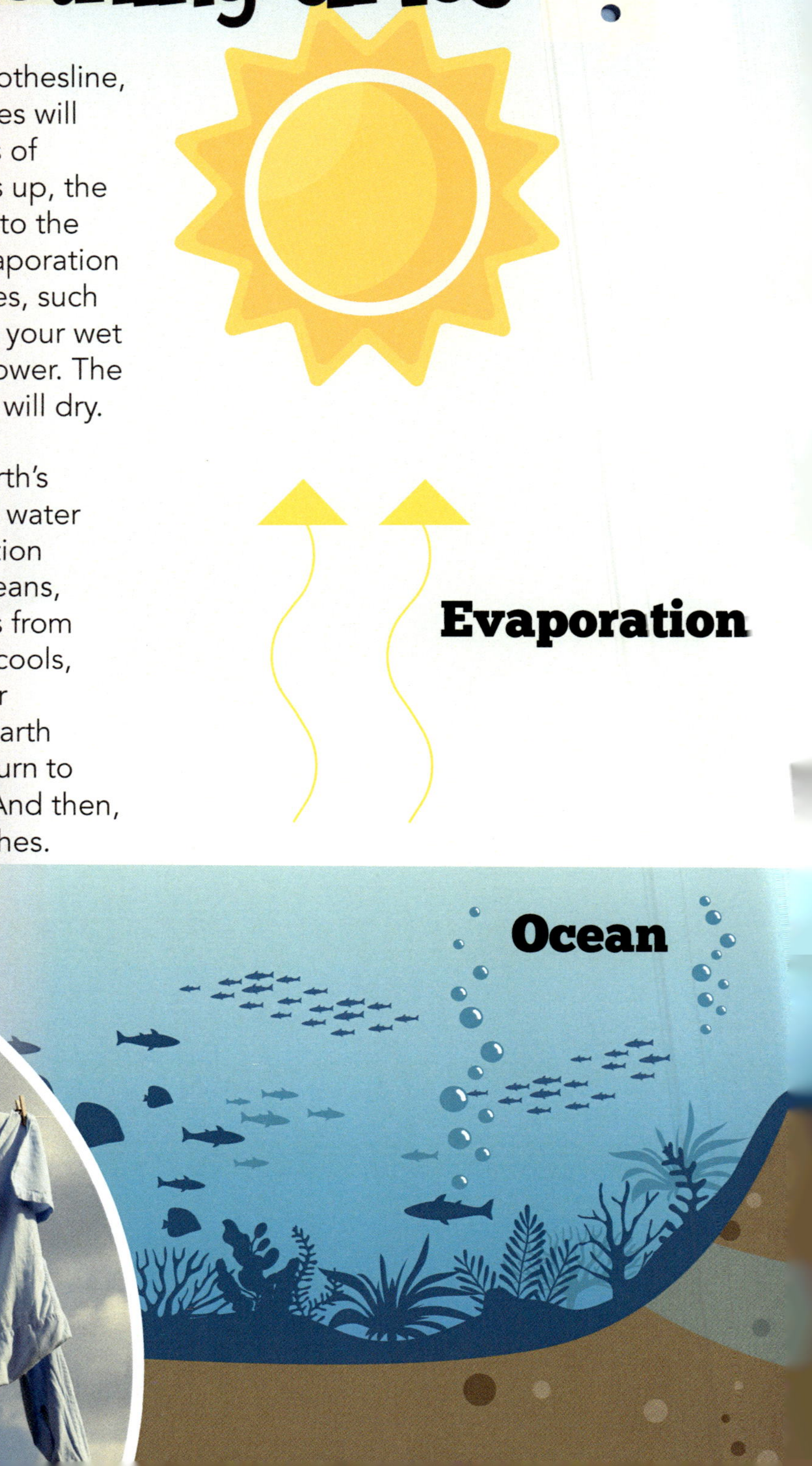

The Water Cycle

Transpiration happens when a plant releases water vapor through its leaves, stems, or other areas.

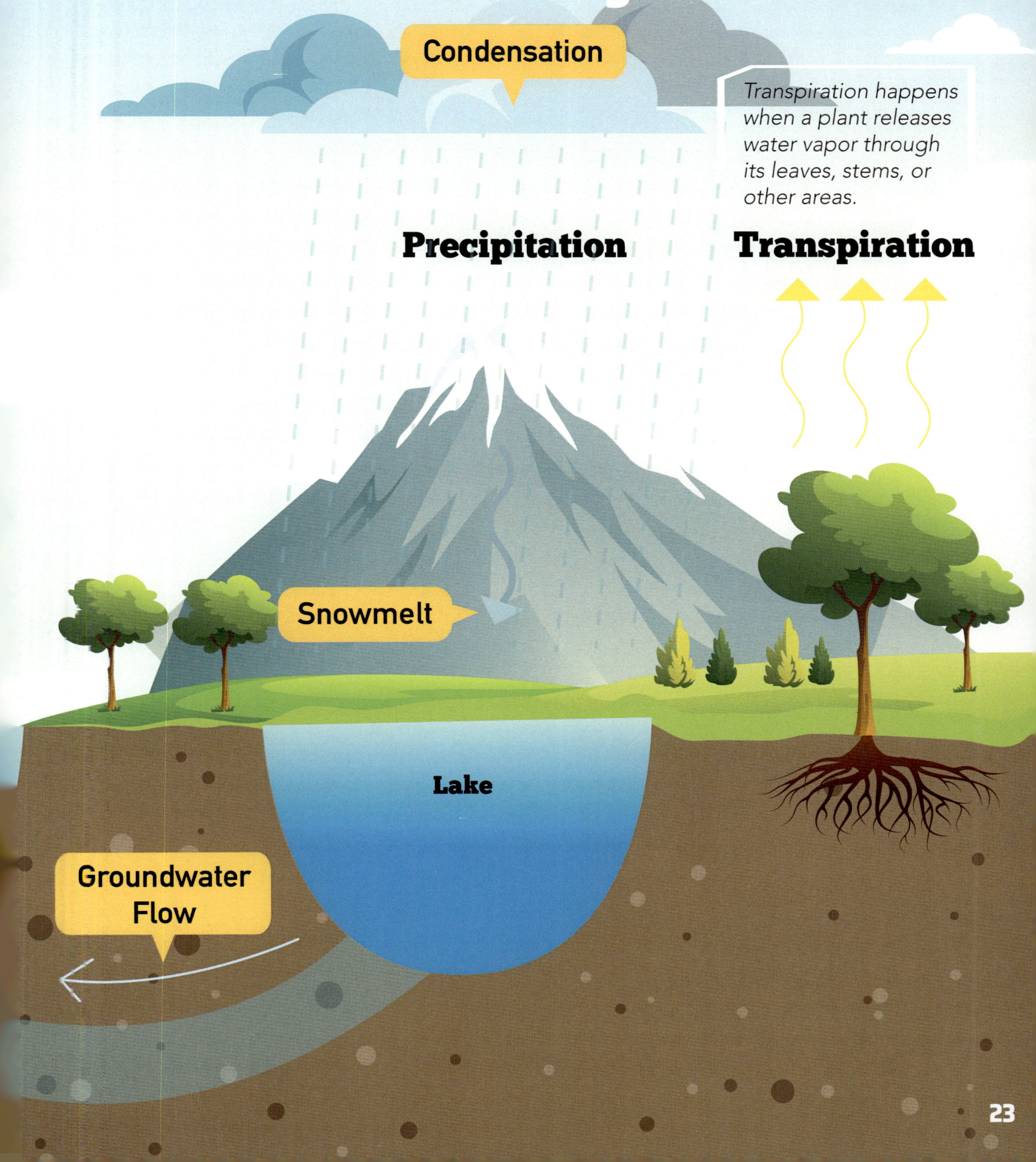

Some liquids, such as milk, juice, or even blood, are mixtures of solids and water. When the water evaporates from these mixtures, the solids are left behind. That's why grape juice and coffee will leave a stain on our clothes. But not all liquids are made of water. Other common liquids are oil and alcohol.

In all liquids, the particles are close together, but not as close together as in a solid. Liquids change their shape based on the shape of their container and they will move to fill the space around them. If you pour a small glass of juice into a bowl, the shape of the liquid will change, but the amount will stay the same. Rain also falls into "containers," such as ponds, lakes, and storm drains.

After a large rainfall, a pond, lake, or storm drain might overflow, causing a flood. In this case, the water will move to fill the space beyond the container.

DID YOU KNOW?

You may have heard incompatible **things or people described as being like oil and water. It's true that water molecules and oil molecules simply do not mix. That's because water molecules are polar: they have a positive** electric charge **at one end and a negative charge at the other. Oil molecules are** non-polar**. That means the two types of molecules will never be attracted to each other.**

Oil molecules are repelled by water molecules. That's why oil floats on water's surface.

What is lava made of?

Lava is **molten** rock that comes through Earth's surface when a volcano erupts. Under the surface, it is known as magma. It might be hard to imagine that rocks can exist naturally in liquid form. But the minerals near Earth's core are so hot that they become liquified! When a volcano erupts, it expels this liquid rock. Once the lava leaves the volcano it cools rapidly. Eventually it becomes a solid again.

We know that a rock sinks in water. It will sink in lava too, but more slowly because lava is denser than water.

A pumice stone—a type of light rock full of holes—comes from lava that cools so quickly, bubbles of gas get trapped.

DID YOU KNOW?

We usually think of metals as solids, but mercury is a metal that is liquid in its natural state. There are also some metals that melt just above room temperature, such as caesium and gallium.

How much of the human body is liquid?

You may have heard that the human body is mostly water—we are in fact 50 to 70 percent water! It may seem hard to believe, since our bodies feel solid. But remember that the most common elements in the body are oxygen, carbon, and hydrogen. A lot of those hydrogen and oxygen atoms make up water molecules. These molecules may not exist as water the way we expect, but all the cells in our body contain water. All the fluids in our body—such as blood and saliva—are water-based. Our organs, tissues, and bones also have water molecules. That is why it's important to stay hydrated—our bodies can't work without water.

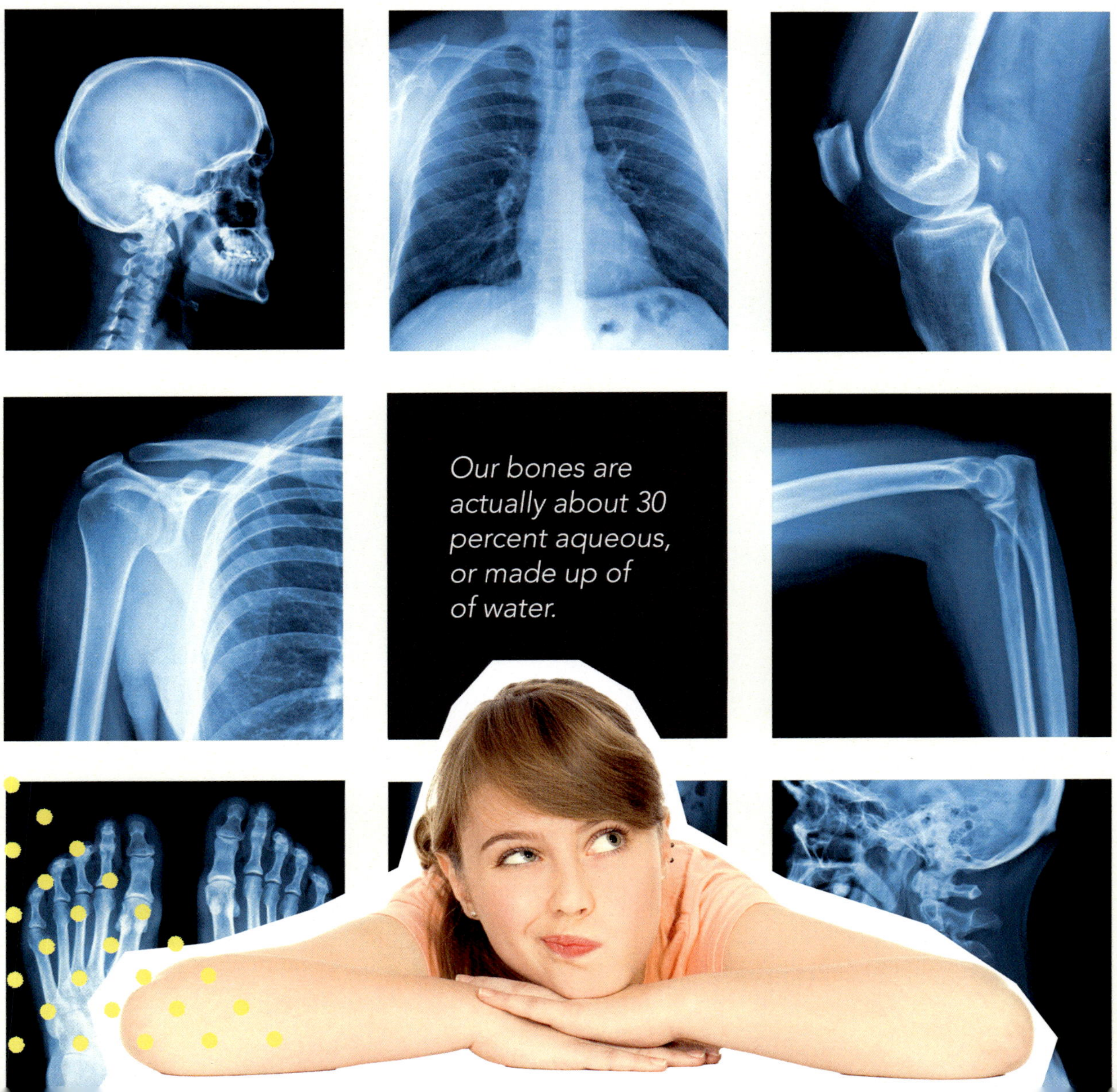

Our bones are actually about 30 percent aqueous, or made up of of water.

What makes a liquid sticky?

Some liquids, such as honey and syrup, are made of sugar molecules **dissolved** in water. When sugar crystals dissolve in water, they make the water more viscous. Viscosity is a measure of how fast or slow a liquid flows. Viscous substances are thick and flow slowly.

As a solid, sugar molecules stick together in crystal form. When water molecules get in between them, the sugar molecules can't stick to each other as tightly. But they still want to stick to something. Water molecules, on the other hand, like to move, but the sugar molecules get in the way and slow them down—making the liquid sticky.

The higher the amount of sugar in the water, the stickier it is.

Is peanut butter a liquid?

Peanut butter might seem like a solid, but it is really a liquid. However, it is different from most liquids. Peanut butter is viscous. Like honey, it flows, and it takes the shape of its container. However, how it flows depends on the force applied to it. A clump of peanut butter on a spoon stays in place, but it can be spread with a knife. Stirring a jar of peanut butter makes it runnier.

In the 1600s, Sir Isaac Newton observed that most fluids have a constant flow. But peanut butter does not. So scientists call liquids like peanut butter "non-Newtonian fluids." These are fluids that don't behave the way Newton predicted.

Heating up peanut butter shows its liquid properties.

What causes wind?

Particles in gases are much farther apart than in liquids. When steam rises from boiling water, for example, gaseous water molecules spread apart and join the gases all around us in the atmosphere. Wind is created when large amounts of gases move quickly around the atmosphere. Like liquids, gases flow and will keep the shape of the container they are in. But unlike liquids, gases have no fixed volume and expand in every direction.

Gases have weight, and as they move, they press against anything they touch. This force is called air pressure. Gases will always move from a high-pressure area, where there are many particles contained closely together, to a low-pressure area, where there is more room for particles to move. If a container of gas is not sealed, the gas will escape. Think about what happens when you blow up a beach ball. If you don't close the cap on the nozzle, the air escapes back into the atmosphere.

Temperature affects air pressure. When gases warm up, they become less dense and rise. The particles move faster and spread out, creating a low-pressure area. As warm air rises, denser, cooler air rushes in to take its place, creating a high-pressure area. Air pressure is highest closest to the ground. Wind is this movement of air from high- to low-pressure areas.

The entire atmosphere is made of gases. About 78% is nitrogen, 21% is oxygen, and 1% is argon. Earth's ***gravity*** *holds these gases in place.*

The bigger the difference between high- and low-pressure areas, the stronger the wind.

The Sun warms up all the particles of gas in the atmosphere. But the atmosphere heats unevenly because different parts of the planet are closer to the Sun, clouds often block the Sun, and at night there is no Sun. This creates different air pressure systems. Storms are caused when large pockets of cold, low-pressure air meet up with large pockets of warmer, high-pressure air. The area where these large pockets of air meet are called fronts.

Over the past 200 years, burning fossil fuels has increased the amount of greenhouse gases, such as carbon dioxide and methane, in the atmosphere. Greenhouse gases trap the Sun's heat. This hasn't changed the oxygen available for us to breathe, but it has influenced the weather. When there are too many greenhouse gases in the atmosphere, Earth's overall temperature increases. This has led to **global warming**, which has increased the number of extreme weather events, such as heat waves and hurricanes.

We need greenhouse gases to make Earth a warm enough place for us to live. But too many of them, such as large amounts of methane from cow burps or carbon dioxide from burning fossil fuels, causes global warming.

Cold fronts are often known for bringing in bad weather, such as heavy rain and thunderstorms.

Burning fossil fuels is a major source of nitrogen dioxide.

Are all gases invisible?

As you look at the air around you, it may seem like all gases are invisible. But there are gases we can see. Nitrogen dioxide is reddish brown, iodine gas is bright purple, and chlorine gas is yellowish green. These gases are very dangerous, so it's a good thing we can clearly see them!

However, invisible gases can also be dangerous. They can be dangerous to breathe, or they can even cause explosions. We use carbon monoxide alarms in our homes to alert us if there is too much of that invisible gas in the air. To help alert people of possible leaks, natural gas companies add a unique smell to the invisible natural gas used to heat homes.

Is fog a gas?

Early morning fog may seem like a gas. It has no shape or volume, and it spreads through the air like other gases. But fog is actually tiny liquid droplets or solid crystals of water suspended in the gaseous air. This is called a colloid. Colloids can also be liquid **solutions**, like milk. When the Sun heats up the fog, the water droplets turn back into gas.

DID YOU KNOW?

Over 100 years ago, the word *smog* was used to describe the real fog that mixed with the smoke produced from burning coal. But today, smog does not require foggy conditions. Hazy air pollution, which we call smog, is made from a photochemical **reaction. The energy of the Sun reacts with gaseous emissions from factories and cars to create a brownish haze in the air.**

Milk is a colloid because it contains small droplets of fat suspended in the liquid.

Smog can be damaging to health as it means there are higher amounts of pollutants in the air.

To steer a hot air balloon, a pilot must use the direction of the wind.

How do hot air balloons fly?

Have you ever noticed that basements feel colder than upper floors? We already know that when gas molecules heat up, they move faster and spread apart. This lowers their density. That's why hot air rises above cold air, making basement air cooler.

Gases have no shape or volume—but they can still be contained. A hot air balloon is a kind of loose container. A heat source at the base of the balloon heats up the air—but instead of floating away, the hot air is trapped in the balloon. Eventually the warm air in the balloon rises above the cooler air outside of it, lifting the balloon's basket off the ground. Turning down the heat in a controlled way cools down the air, slowly lowering the balloon to the ground.

Why do helium balloons float?

When we blow up regular balloons, they float lightly to the ground. The gas in our lungs has a similar density as the air around us. But helium is the second element on the periodic table, which means it is much lighter than the gases in the atmosphere. When helium gas is contained in a balloon, it will rise higher and higher and float away.

A helium balloon will eventually deflate and sink as helium gas slowly escapes through the balloon.

DID YOU KNOW?

Large volumes of gas can be compressed in special containers. This pushes the molecules together, often into a liquid form. Then, small amounts of gas can be released for different purposes. You may see compressed gas in the propane tanks for barbecues or the helium tanks for balloons.

Compressed gas is also used in aerosol or spray cans for products such as hair spray, sunscreen, and air freshener.

CHAPTER 5: QUESTIONS ABOUT CHANGING STATES OF MATTER

How do melting icebergs affect sea levels?

We know that matter doesn't disappear—it just changes state. If we heat up water, it turns to gas. If we freeze water, it becomes ice. These changes happen every year at Earth's North and South Poles. Sea ice melts as the temperature rises in summer, then refreezes during the long winter months. But the amount of water in the oceans doesn't change. It simply changes state as part of the water cycle.

Icebergs, however, are different from sea ice. Icebergs are large masses of ice that break away from glaciers and **ice sheets** that cover the land near the poles. As Earth's overall temperature rises due to global warming, more and more glaciers are breaking and melting, and more icebergs are added into the oceans. Because this solid water was previously covering the ground, it wasn't part of the overall volume of Earth's oceans. So when icebergs break away from land ice and melt, the amount of water in the oceans increases and the sea level, or surface of the ocean, rises. Scientists estimate that if all the ice covering Greenland melted, the global sea level would rise by about 23 feet (7 m).

Most icebergs are 1,000 to 1,600 feet (300 to 500 m) wide, but some can be wider than 0.6 miles (1 km)!

The Greenland ice sheet covers about 80 percent of the island, or 660,000 square miles (1,710,000 sq. km).

Melting sea ice may not change the volume of Earth's oceans, but it does change their temperature. Without sea ice to reflect the Sun's rays, the oceans absorb more sunlight and become warmer. The temperature of the air also stays warmer without massive amounts of ice at the poles. This leads to more evaporation from the oceans and more water vapor in the air. All of these changes in the gases of the atmosphere have created more extreme and unpredictable weather around the world.

The temperature affects matter all around us. Freezing, melting, and evaporating are three ways we see changing states of matter due to temperature. But not all liquids freeze, melt, or evaporate the same way. For example, alcohol freezes or solidifies at a much lower temperature than water, and will evaporate at a lower temperature. And some oils will become solid at higher temperatures than water and don't evaporate when heated in a normal way—that is why oils are used in cooking.

Helium would only become solid in an atmospheric pressure 20 times that of Earth.

DID YOU KNOW?

We already know that a balloon filled with helium floats. But helium has another interesting property: it is the only liquid that does not freeze. Helium gas becomes liquid at the extremely low temperature of –452.2 degrees Fahrenheit (–269 degrees Celsius). However, at these low temperatures, its atoms continue to vibrate, which creates enough energy to keep it from freezing solid.

Hurricanes need warm ocean waters and a lot of moisture in the air. Warmer ocean temperatures and higher amounts of evaporation have increased the number and strength of hurricanes.

Why does a cold drink sweat in hot weather?

When a cold drink is surrounded by warm air, water vapor in the air starts to cool down and water droplets form on the outside of the glass or bottle. This change of state from gas to liquid is an example of condensation. Condensation also fogs up the bathroom mirror when someone takes a hot shower. The dew that forms on the grass in the early morning is also caused by condensation. The ground cooling at night causes water vapor in the air to condense.

The steam from the hot shower creates small droplets of water on the cooler bathroom mirror.

Condensation is what makes eyeglasses foggy when someone goes from the cold outdoors to a warm indoors.

What makes the windows frosty?

Frost is similar to dew, but has an important difference: it happens when water vapor turns directly into a solid, skipping the liquid state. This is a process known as deposition. In cold weather, the water vapor inside a warm home cools down very quickly near a cold window. This turns the water from its gaseous form into a solid, forming frosty crystals on the glass.

DID YOU KNOW?

The opposite of deposition is called sublimation—when a solid skips the liquid form and changes into a gaseous state. Dry ice is frozen carbon dioxide. As dry ice "melts," or sublimates into a gas, it looks like fog. Performers often use dry ice to create the effect of fog on stage. Dry ice can also be used to keep food cold without a refrigerator.

Adding water to dry ice will speed up the rate of sublimation.

Why does a soda can explode in the freezer?

Most substances decrease in volume when they change state from liquid to solid. That's because their molecules get closer together. But water is unique. It expands when it freezes. If you put a container of water in the freezer without room for the water to expand, the container is likely to crack.

Carbonated drinks, such as soda, contain little bubbles of carbon dioxide gas. As the soda freezes and expands, it pushes the gas to the edge of the can. This puts pressure on the gas, which forces the container to crack open and spew the freezing liquid all over the freezer!

Sea ice floats on the surface of an ocean because water expands in its solid form—making the ice less dense than the liquid water below.

What happens to food after it is put in the compost?

After time, the food scraps in a composter start to rot or biodegrade. To biodegrade is to break down. It may seem like the food is breaking down into nothing, but remember that matter cannot be created or destroyed, only transformed. The food scraps actually turn into a soil-like substance.

Composting is part of the carbon cycle. Bacteria eat away at the food scraps. The bacteria's waste puts **nutrients** into the compost and carbon gases into the air. Then the compost can be added to soil, where plants can use the nutrients and carbon in the air for photosynthesis.

DID YOU KNOW?

Plastic is not biodegradable. It can break into smaller and smaller pieces, but it cannot become part of the carbon cycle. Plastic must be melted to liquid plastic and reformed into another solid before it can be reused. Some plastic breaks down into tiny pieces called microplastics.

Scientists have found evidence that microplastics are present in clouds, showing that they have polluted the water cycle.

Glossary

atmosphere (AT-muhs-feer): The layers of gases surrounding a planet

carbon (KAHR-buhn): A non-metal element present in all living things

chemical reactions (KEM-i-kuhl ree-AK-shuhns): Processes by which substances react with each other to create new substances

condensation (kon-den-SEY-shuhn): The change of state from gas to liquid

crystalline structure (KRIS-tl-ahyn STRUHK-cher): An arrangement of atoms in a three-dimensional, repeated pattern

dense (dens): Having a lot of matter packed closely together

dissolved (dih-ZOLVD): Of a solid, mixed into a liquid to become part of it

electric charge (ih-LEK-trik chahrj): A property of matter that occurs when the number of protons and electrons differ in an atom, causing objects to attract or repel each other

evaporation (ih-vap-uh-REY-shuhn): The change of state from liquid to gas

force (fawrs): A push or pull that causes a change in movement

fossil fuels (FOS-uhl FYOO-uhls): Fuels, such as coal, made from ancient plant or animal remains

global warming (GLOH-buhl wawrm-ing): The gradual increase in Earth's average temperature, caused by human activity

gravity (GRAV-i-tee): The force that pulls objects toward Earth's center, or that pulls two objects with mass toward each other

groundwater (GROUND-waw-ter): Water held underground in the spaces in soil, rock, and sand

ice sheets (ahys sheets): Masses of ice larger than 20,000 square miles (50,000 sq. km) that cover huge areas of land

incompatible (in-kuhm-PAT-uh-buhl): Not able to work well together

Indigenous (in-DIJ-uh-nuhs): The first peoples to live in an area

insulates (in-suh-LEYTS): Covers something to protect it or stop heat or sound from escaping

mass (mas): A measure of the amount of matter in an object

molten (MOHL-tn): Became liquid by heat

newtons (NOOT-ns): A measure of force

non-polar (non-POH-ler): Molecules that do not have opposite positive or negative poles

nutrients (NOO-tree-uhnts): Substances used by living things to live and grow

particles (PAHR-ti-kuhls): Small pieces of matter

photochemical (FOH-toh-KEM-i-kuhl): A chemical action triggered by light

photosynthesis (foh-tuh-SIN-thuh-sis): The process in which plants use the Sun's energy to create carbon and oxygen

porous (PAWR-uhs): Having small holes through which liquids or gases can pass through

proteins (PROH-teens): Substances found in living organisms that do important work in the body's cells

soda ash (SOH-duh ash): A powdery white substance made from sodium carbonate

solutions (suh-LOO-shuhns): Two or more substances that have mixed completely

suspended (suh-SPEN-did): Of small pieces of matter, spread within a fluid

water cycle (WAW-ter SAHY-kuhl): The continuous movement of water on Earth and in its atmosphere

wavelengths (WEYV-lengkths): The distance of a sound or light wave

Index

air pressure 30–32, 40
atmosphere 22, 30, 32, 37, 40
atoms 4, 6–8, 10, 13, 15, 16, 27, 40

Big Bang 6–7

carbon 8–9, 10, 13, 16, 27, 32, 34, 43, 45
carbon cycle 9, 45
carbon dioxide 8–9, 32, 43–44
condensation 22, 42
crystals 15, 20–21, 28, 35, 43

density 4, 6, 12, 14–16, 26, 30, 36–37, 44
deposition 43

elements 8, 10, 13, 16, 27, 37
evaporation 22, 24, 38, 40–41
extreme weather 32–33, 40–41

gases 4, 6, 8–9, 13, 26, 30, 32, 34–38, 40, 42–45
global warming 32, 38
gravity 12, 30

helium 37, 40
hydrogen 8, 10, 13, 15, 27

ice 5, 8, 15, 16, 20, 38–40, 43, 44

Law of Conservation of Matter 6
liquids 4, 6, 8, 13, 15, 18–22, 24, 26–30, 35, 37, 40, 42–45

mass 4, 10, 12
Mohs scale 16
molecules 8, 13, 15, 16, 18, 20–22, 25, 27, 28, 30, 36–37, 44

Newton, Isaac 29
nitrogen 13, 30, 34

oxygen 8, 13, 27, 30, 32

particles 4, 6, 8, 10, 12, 14, 16, 18, 24, 30, 32
periodic table 10, 37
photosynthesis 9, 45
plastic 19, 45

snow 20, 22
solids 4, 6, 8, 13–16, 18–21, 24, 26–29, 35, 38, 40, 43–45
sublimation 43

universe 4, 6, 12

viscosity 28–29
volume 4, 30, 35–38, 40, 44

water 5, 6, 8–9, 11, 14–18, 20, 22–28, 30, 35, 38, 40–45
water cycle 8, 22–23, 38, 45

Comprehension Questions

1. **Scientists believe _______ are the smallest pieces of matter.**
   a) electrons
   b) quarks
   c) atoms

2. **True or False:** Sublimation and deposition are opposite processes.

3. **_________ do not have volume.**
   a) Solids
   b) Liquids
   c) Gases

4. **True or False:** The human body is entirely made up of oxygen, hydrogen, carbon, and nitrogen.

5. **_________ is the process by which liquids become gases.**
   a) Condensation
   b) Evaporation
   c) Transpiration

6. **True or False:** A mineral's hardness is measured in newtons.

7. **A _______ solid will break when it is bent or pulled.**
   a) brittle
   b) elastic
   c) ductile

8. **True or False:** Peanut butter is an amorphous solid.

1. b, 2. True, 3. c, 4. False, 5. b, 6. False, 7. a, 8. False

About the Author

Rachel Stuckey is a writer, editor, and learning content designer from Toronto. She has spent much of her life as a digital nomad, working remotely as she travels the world. She has written more than 30 books for young readers.

Crabtree Publishing

crabtreebooks.com 800-387-7650

Hardcover 978-1-0398-8094-8
Paperback 978-1-0398-8455-7
Ebook (pdf) 978-1-0398-8214-0
Epub 978-1-0398-8334-5

Printed in the U.S.A./CP2025

Published in Canada
Crabtree Publishing
616 Welland Avenue
St. Catharines, Ontario
L2M 5V6

Published in the United States
Crabtree Publishing
347 Fifth Avenue
Suite 1402-145
New York, NY 10016

Written by: Rachel Stuckey
Series development: Janine Deschenes
Editorial coordinator: Kathy Middleton
Editor: Janine Deschenes
Proofreader: Melissa Boyce
Art direction and layout: Katherine Kantor and Margaret Salter
Photo research: Katherine Kantor
Production manager: Candice Campbell

Images:
NASA: JSC: p. 12
Shutterstock: p. 44 (top right)
Wikimedia Commons: Jurii: p. 3 (center)
All other images: Shutterstock

Library and Archives Canada Cataloguing in Publication
Available at the Library and Archives Canada

Library of Congress Cataloging-in-Publication Data
Available at the Library of Congress